EPIC ANIMAL JOURNEYS

Unusual Animal Journeys

NATALIE HYDE

CRABTREE
PUBLISHING COMPANY
WWW.CRABTREEBOOKS.COM

Author:
Natalie Hyde
Series research and development:
Reagan Miller, Janine Deschenes
Editorial director:
Kathy Middleton
Editor:
Sonya Newland, Ellen Rodger
Proofreader:
Izzi Howell, Melissa Boyce
Graphic design:
White-Thomson Publishing
Katherine Berti
Image research:
Natalie Hyde
Print and production coordinator:
Katherine Berti

Images:
Alamy
Jim West p. 21
iStock
sarkophoto p. 5t; wrangel p. 9t; Vassiliy Vishnevskiy p. 23t; Andrew Banham p. 25m; Leon Kiss p. 27b; JohnCarnemolla p. 28b;
Shutterstock
John Orsbun p. 13b;
Getty Images
AFP p. 9b; Lucia Teru p. 13t; Bruce Lichtenberger p 15t; John E Marriott p. 15b; Paul Kay p. 17t; Handout p. 19t;
Wikimedia Commons
CSIRO_ScienceImage p. 8t

All other images from Shutterstock

Contents page: Gray squirrel

Library and Archives Canada Cataloguing in Publication

Title: Unusual animal journeys / Natalie Hyde.
Names: Hyde, Natalie, 1963- author.
Description: Series statement: Epic animal journeys | Includes index.
Identifiers: Canadiana (print) 2019009799X | Canadiana (ebook) 20190098007 |
ISBN 9780778763703 (hardcover) |
ISBN 9780778763789 (softcover) |
ISBN 9781427123510 (HTML)
Subjects: LCSH: Animal migration—Juvenile literature. |
LCSH: Nature—Effect of human beings on—Juvenile literature.
Classification: LCC QL754 .H932 2019 | DDC j591.56/8—dc23

Library of Congress Cataloging-in-Publication Data

Names: Hyde, Natalie, 1963- author.
Title: Unusual animal journeys / Natalie Hyde.
Description: New York, New York : Crabtree Publishing Company, [2019] |
Series: Epic animal journeys | Includes index.
Identifiers: LCCN 2019015872 (print) | LCCN 2019018067 (ebook) |
ISBN 9781427123510 (Electronic) |
ISBN 9780778763703 (hardcover) |
ISBN 9780778763789 (pbk.)
Subjects: LCSH: Introduced organisms--Juvenile literature. |
Nature--Effect of human beings on--Juvenile literature.
Classification: LCC QH353 (ebook) | LCC QH353 .H93 2019 (print) |
DDC 591.6/2--dc23
LC record available at https://lccn.loc.gov/2019015872

Crabtree Publishing Company
www.crabtreebooks.com 1-800-387-7650

Printed in the U.S.A./082019/CG20190712

Published in Canada
Crabtree Publishing
616 Welland Ave.
St. Catharines, Ontario
L2M 5V6

Published in the United States
Crabtree Publishing
PMB 59051
350 Fifth Avenue, 59th Floor
New York, New York 10118

Published in the United Kingdom
Crabtree Publishing
Maritime House
Basin Road North, Hove
BN41 1WR

Published in Australia
Crabtree Publishing
Unit 3 – 5 Currumbin Court
Capalaba
QLD 4157

Contents

Introduction

On the Move

Migrations are regular journeys made by animals. Animals migrate for different reasons. Some are looking for food or shelter. Others find a mate and have babies in special areas. Still others look for warmer or cooler temperatures when the seasons change. Soldier crabs in the Caribbean migrate from land to the shore to lay eggs. In Norway, rodents called lemmings migrate to find food. Dragonflies follow the rains from India to Africa so they can breed in freshwater. Snow geese migrate from Texas and Mexico to the Arctic to nest.

It is safer for animals to migrate in a large group because it gives them some protection from predators.

Animals may have trouble finding food or a mate if human-made structures block their way.

Unusual Journeys

Some animal movements are not part of a regular migration. They may be accidental, unusual, or caused by environmental changes. Insects or small animals can be hiding in shipments from other countries. Birds can fly in the opposite direction from the rest of their flock. Pipelines can block normal migration routes, forcing animals to take a new path.

Climate change is forcing tropical fish like the blue-barred parrotfish toward the poles, where the sea is colder.

The Changing Planet

Climate change is another reason why animals make these unusual journeys. They may be used to warmer or cooler temperatures and have to move to a new location as weather patterns change. Not enough rain—or too much rain—can ruin a particular food source for a species. They may need to migrate to find a new source. When this happens, we may see animals in areas they have never been before. This can alter the **ecosystem**, which then puts other animals at risk.

Irruptions

Irruptions are irregular movements of large numbers of animals to areas where they normally aren't found. Irruptions often happen because of a food shortage after a population explosion. Hawaiian honeycreeper birds are a kind of finch. They are related to rosefinches, which are not native to the Hawaiian islands. These birds are known to irrupt when food sources get low. Scientists believe that some rosefinches may have been carried away by a storm during an irruption and landed on the islands of Hawaii.

Scientists believe the first rosefinch ancestors of the honeycreepers arrived in the Hawaiian islands between 5.8 and 7.2 million years ago.

Invasive Species

Invasive species are animals that arrive in a new **habitat** by accident, often due to human activity. These animals may be living in fresh food picked and shipped to other countries. They can also be hidden among packing material around products sent across the world. Once in a new location, they reproduce and make a new colony. They are called "invasive" because they affect or invade the local food chain. They compete for food and shelter. Sometimes they wipe out other species or bring new diseases. They take over the habitat of other native species.

Spiders like this hunstman sometimes hitch a ride to new countries in crates of fruit.

Vagrancies

Vagrancies are animals that end up far outside their normal range. They may be blown off course during storms, ride ocean currents, or accidentally hitch a ride on ships or planes. During a cyclone called Storm Eva, a loggerhead sea turtle was forced from the Mediterranean to the shores of Ireland. Loggerheads usually live in the Mediterranean Sea.

Baby loggerhead sea turtles make their first migration right after birth. The 8,000-mile (12,875 km) migration around the Atlantic Ocean can take six to 12 years.

Australian Plague Locust

Habitat	Grasslands of Australia
Known for	Huge swarms of more than 1 billion locusts
Journey type	Irruption

When locusts are airborne, a huge gust of wind or a storm could blow them out to sea.

Flight in the Night

Locusts lay their eggs in warm, wet areas with plenty of new grass for young **nymphs** to feed on. When strong, warm winds and rain start to form, locusts begin to gather in a large swarm. They follow these rains to find a good area to lay the next generation of eggs. As night falls, the locusts take off into the wind. They use the rising warm air to fly as high as 3,200 feet (975 m).

Locusts can be carried by the wind and travel hundreds of miles in a single night. Sometimes they travel for several days and end up infesting a place that previously had few or no locusts. This event is known as an outbreak, or **plague**.

Locusts usually like to live alone. Before swarming, a chemical in a locust's brain makes it want to join together with other locusts.

Sometimes, sporting events have to be canceled when swarming locusts cover fields and paths, making it dangerous for athletes.

Plague Problems

Locust swarms damage grasslands and crops. If they land at night, they can eat an entire crop or field before daylight. They will feed on fruit trees, vines, shrubs, and lawns. Buildings can be completely covered. People have trouble driving because locusts cover their windshields!

Gray Squirrel

Habitat	Hardwood forests of North America
Known for	Huge squirrel migration of 1968
Journey type	Irruption

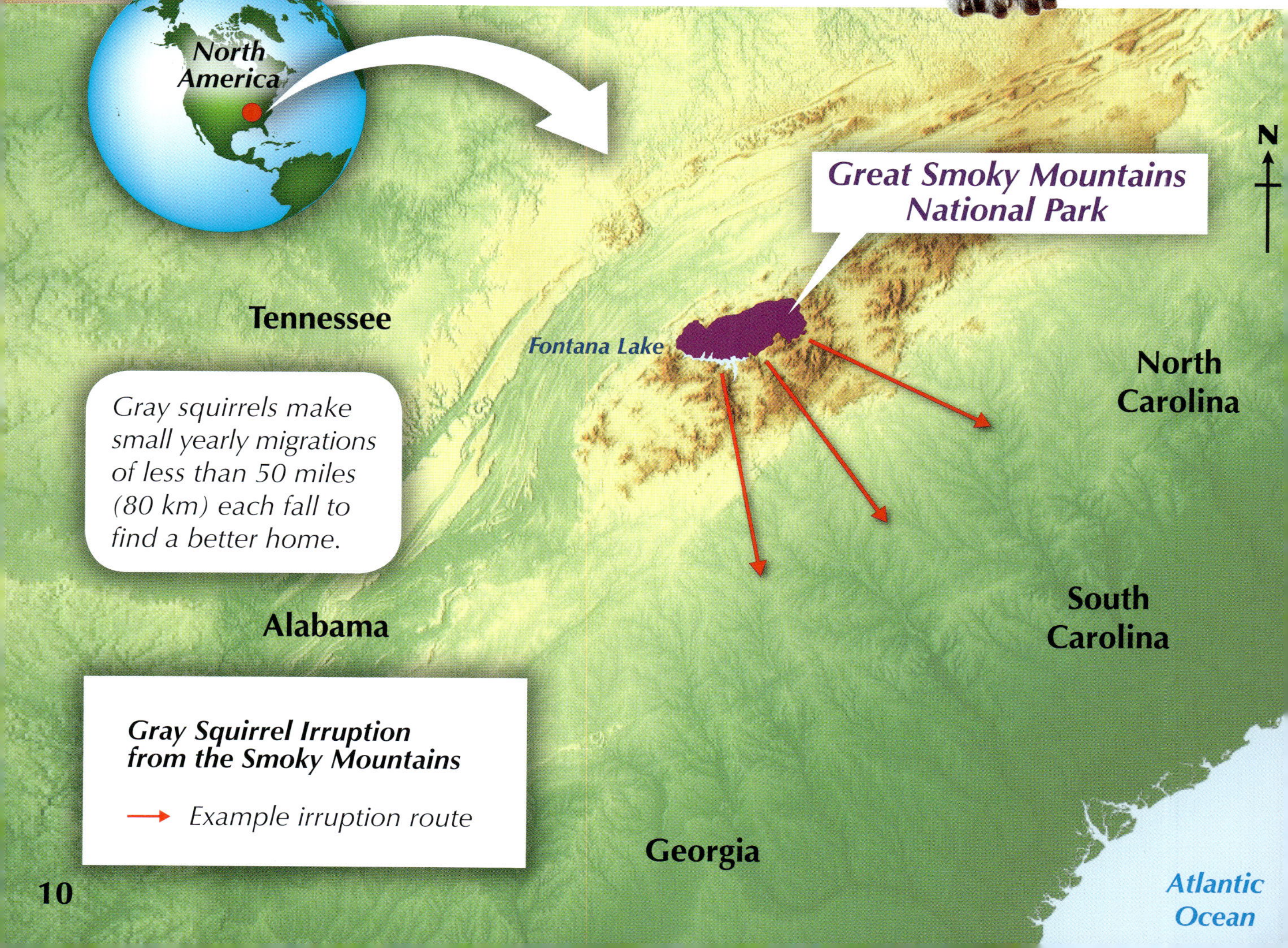

Gray squirrels make small yearly migrations of less than 50 miles (80 km) each fall to find a better home.

Gray Squirrel Irruption from the Smoky Mountains

→ *Example irruption route*

On the Move

Gray squirrels do not regularly migrate. However, in 1968, eastern gray squirrels went on the move. They headed from the Great Smoky Mountains across areas that included Tennessee, North Carolina, Maryland, and New York. Thousands died when they were hit by cars or drowned as they crossed rivers.

When migrating squirrels arrive in a new area, they are in danger from hawks, owls, dogs, cats, and sometimes humans, who hunt them for food, their tails, or for sport.

Past and Future

The year before there had been a bumper crop of acorns. This increase in food resulted in a population explosion of baby squirrels. By the time these babies grew up, crop levels had returned to normal. The squirrels had to go in search of new food sources.

This had happened many times before. In 1857, a cold snap killed off the buds of nut trees. This is a major food source for squirrels. The squirrels had to migrate in search of food. People saw squirrels crossing roads in the thousands. It has also happened several times since, although not on such a large scale.

Barred Knifejaw

Habitat	Waters off the coast of Japan, Korea, and China
Known for	Hitchhiking to the west coast of North America
Journey type	Vagrancy

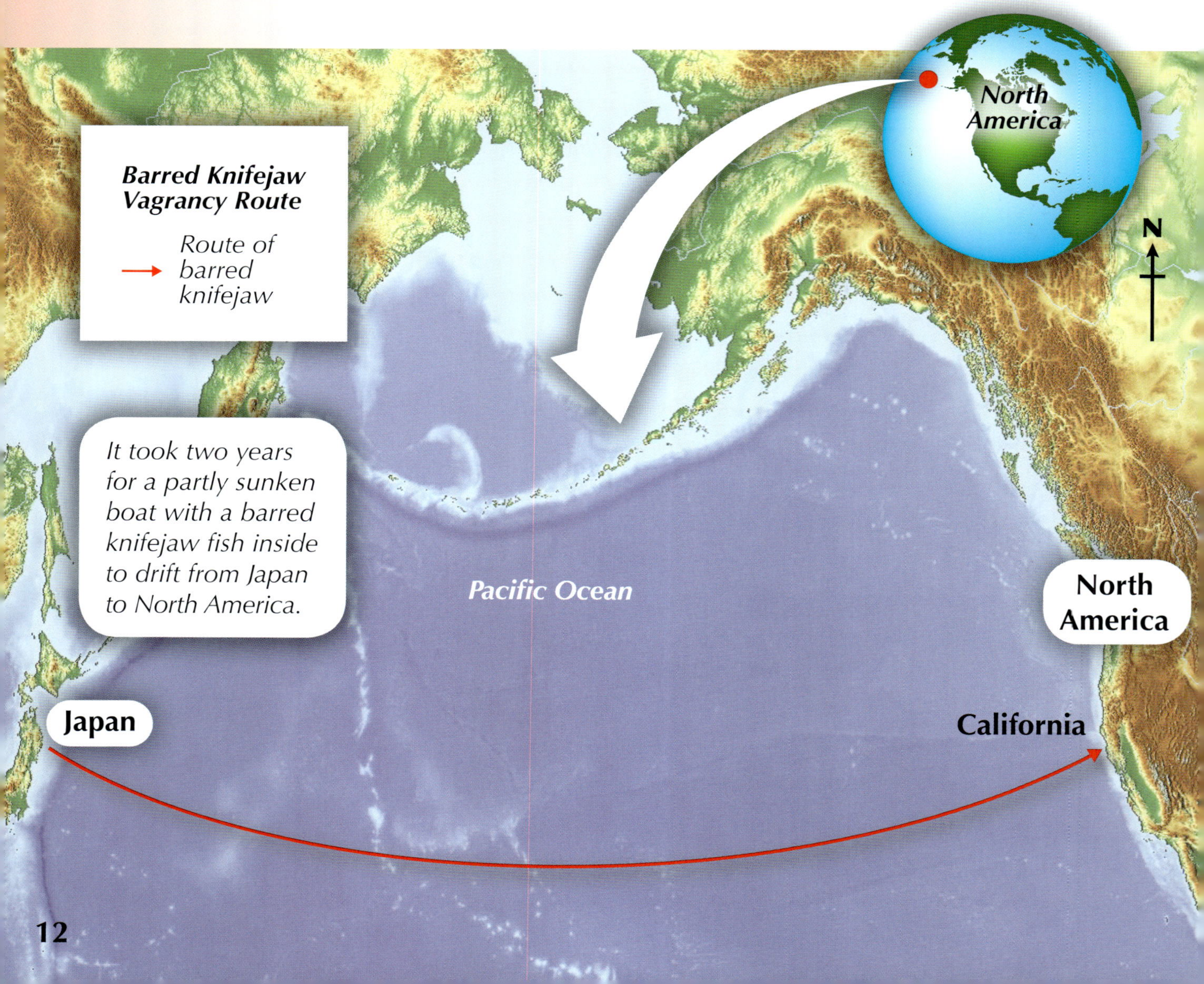

Floating Across the Pacific

In December 2014, a diver near Monterey, California, saw a barred knifejaw. These fish live among the rocky coral reefs of the northwest Pacific Ocean. No one could figure out how it got to California. Since then, divers have seen four or five more of these fish off the coast of California.

On March 11, 2011, a large wave called a tsunami hit Japan. It sent boats, ships, and tons of other debris out into the Pacific Ocean. Some fish and other marine animals became trapped inside these vessels, along with marine plants such as seaweed. Having this food source allowed the animals inside to survive the long journey across the ocean. The slow speed of travel gave them time to adjust to different water temperatures.

Cool Waters

These barred knifejaws have earned the nickname "tsunami fish." Several adult knifejaws have been seen off the coast of California. Although they are mature enough to reproduce, they may not do so. The cooler water temperatures around the coast of California might mean that the eggs can't hatch or young can't survive.

The barred knifejaw is also known as the striped beakfish. It lives in rocky reefs.

Damaged boats can become life rafts not only for fish, but also for plants and crustaceans, bringing them to new habitats.

Snowy Owl

Habitat	**Tundra** for breeding, prairies and the Great Lakes during fall and winter
Known for	Following Arctic lemming migrations
Journey type	Irruption

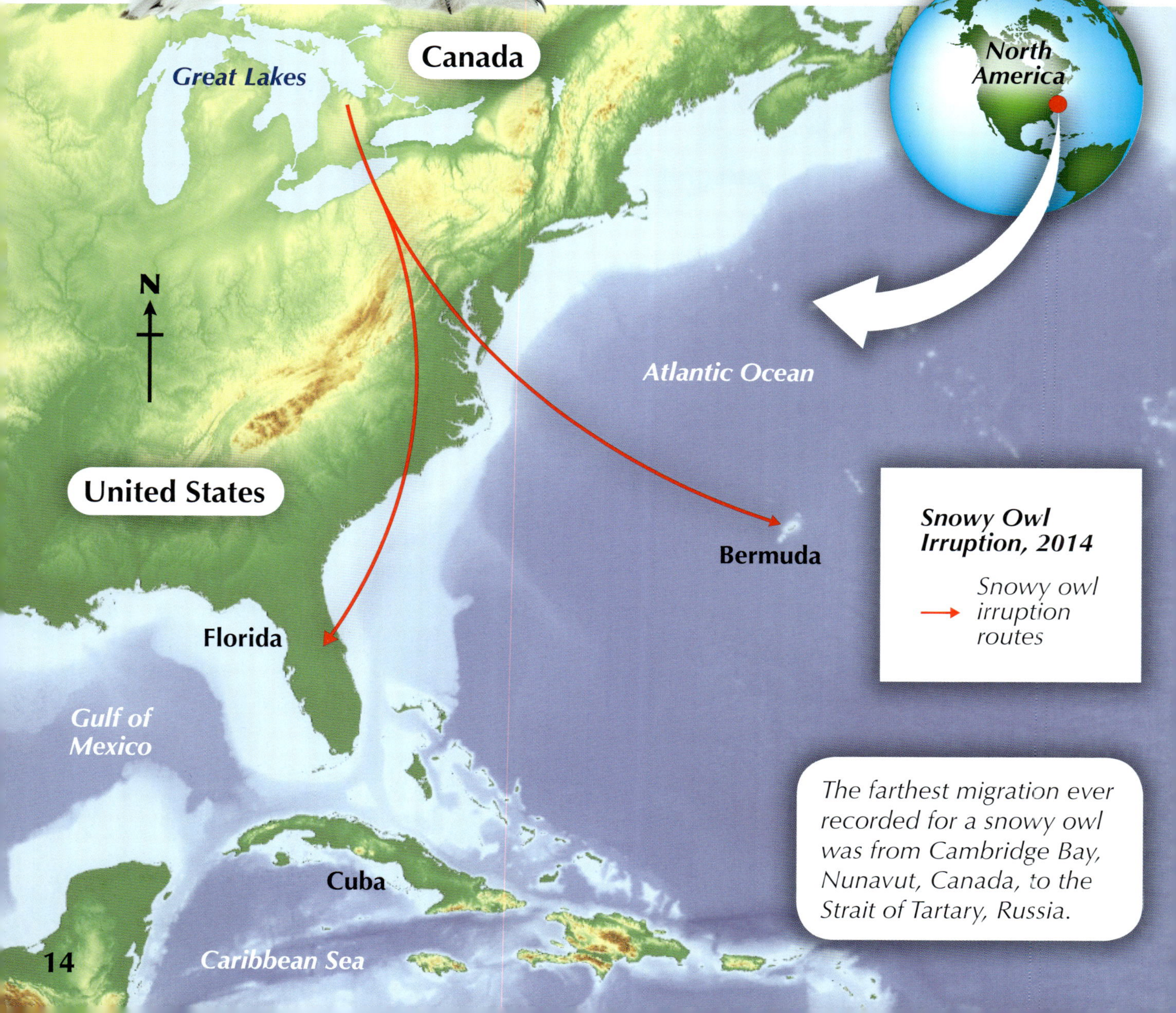

The farthest migration ever recorded for a snowy owl was from Cambridge Bay, Nunavut, Canada, to the Strait of Tartary, Russia.

Risky Migration

Snowy owls normally migrate between the Arctic, where they have their **breeding grounds**, and the prairies and Great Lakes region. But sometimes, a massive snowy owl irruption occurs. The owls show up much farther south than normal. Many people think this irruption is due to low numbers of the owl's **prey**, such as Arctic lemmings and mice. They believe that the birds are migrating to find food sources. In fact, this is not true. Scientists do not fully understand why these irruptions happen.

Irruption migration groups are mostly made up of younger snowy owls.

Too Much Food?

The snowy owl irruption of 2014 came after a year when there was a huge increase in the number of lemmings. With so much food to hunt, snowy owls hatched a record number of eggs. The next spring, all those young snowy owls headed south. They were sighted as far south as the Caribbean Sea.

Arctic lemming

Not all irruptive snowy owls survive. Some collide with vehicles. Others eat poison laid for rodents or are electrocuted by power lines on the way.

Small irruptions happen every four or five years. But a mega-irruption, like the one from 2013 to 2014, happens once or twice in a lifetime.

Snowy owls do not usually irrupt in large groups, but smaller groups may band together to head in the same direction.

European Green Crab

Habitat	Native to the northeast Atlantic Ocean and Baltic Sea
Known for	Being the "World's Worst Alien Invasive Species"
Journey type	Invasive species

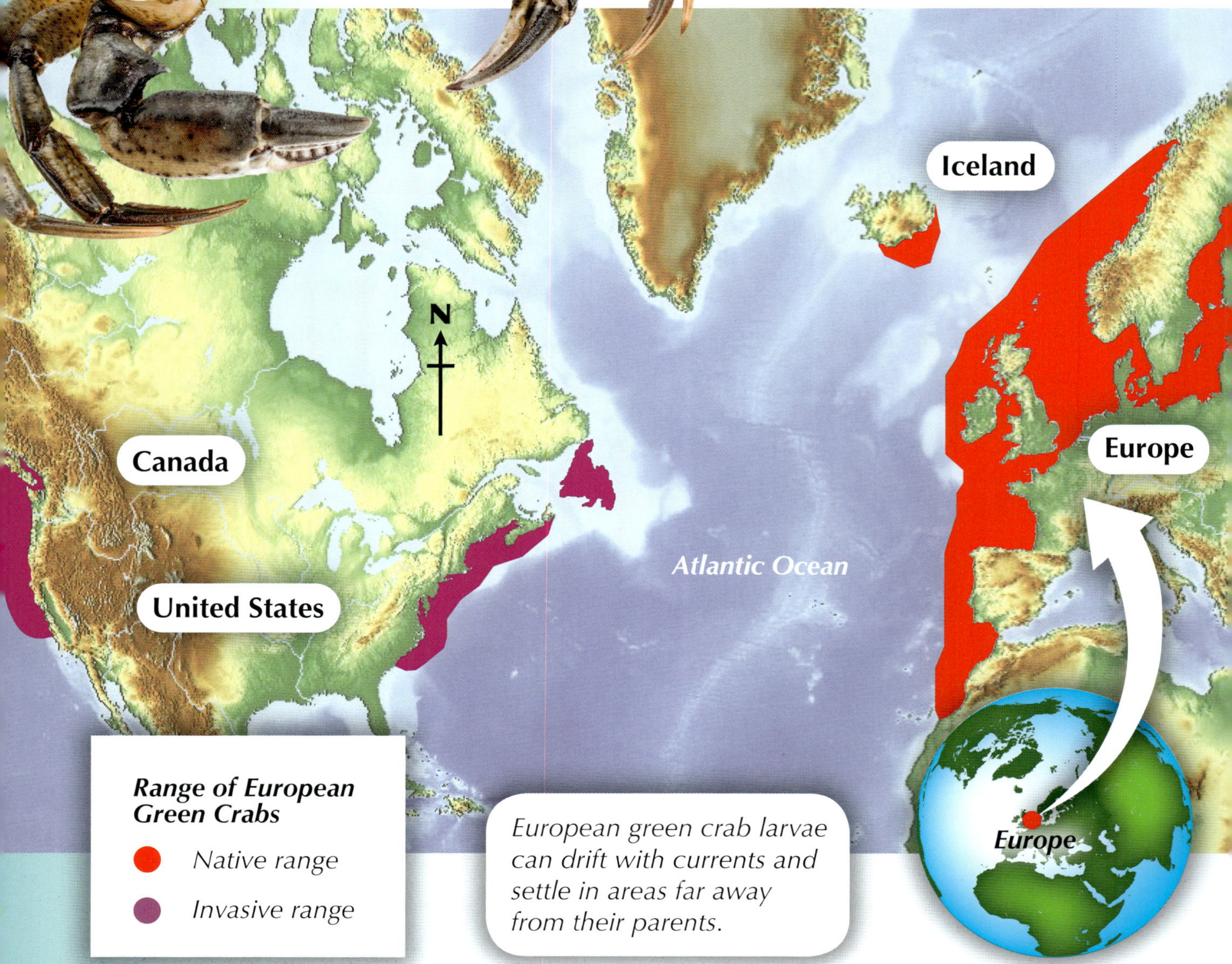

European green crab larvae can drift with currents and settle in areas far away from their parents.

Moving In and Taking Over

European green crabs are native to the coast of Europe on the northeast Atlantic. They can live in many different areas and in different water temperatures. Because of this, these crabs have become an invasive species in many different parts of the world.

The east coast of North America became a new home for green crabs in 1817. They may have been brought over accidentally on sailing ships in packing material or in seaweed. Now they can be found all the way up and down the coast from Newfoundland to South Carolina.

In 1989, people noticed that green crabs had arrived in San Francisco Bay. They probably made this journey in **ballast water**. The ships take the water on before they sail and let it go when they dock.

Female green crabs can release up to 185,000 eggs once or twice a year.

The European green crab is so good at hunting that it leaves little food for native crabs.

Crab Control

The Green Crab Project was set up to try and control the crabs on the west coast. Scientists removed many adults, thinking this would mean fewer babies. But instead, the population skyrocketed. The adults had been eating baby green crabs, keeping them in check! Scientists are looking for new ways to control crab numbers.

Saltwater Crocodile

Habitat	Eastern India, Southeast Asia, and northern Australia
Known for	Riding ocean currents for hundreds of miles
Journey type	Vagrancy

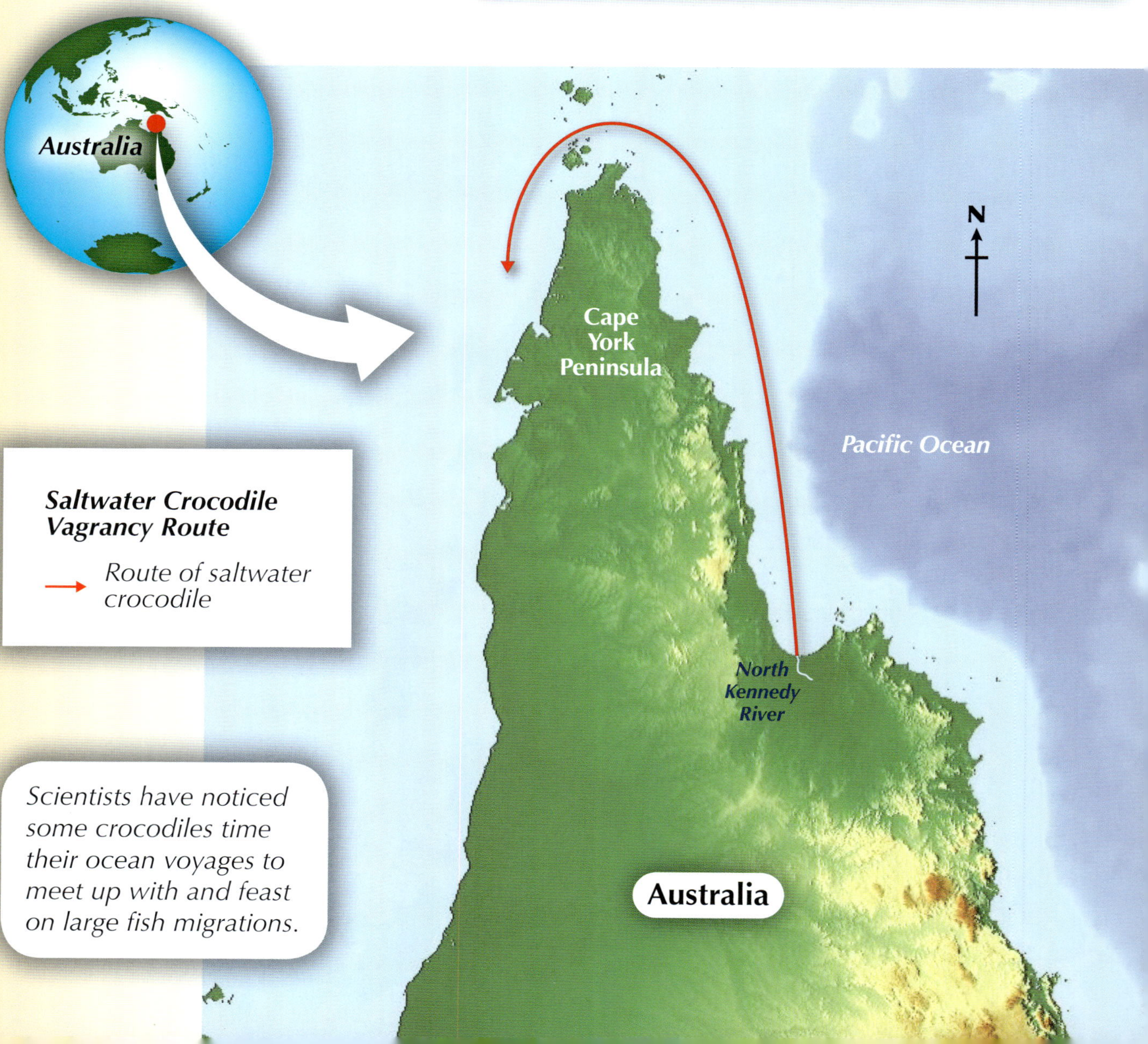

Scientists have noticed some crocodiles time their ocean voyages to meet up with and feast on large fish migrations.

Crocodiles Everywhere!

"Salties" actually live in freshwater rivers or deltas—areas where rivers meet the ocean. They eat and drink on land. These crocodiles have a huge range and were found on islands in Southeast Asia and northern Australia, hundreds of miles apart. No one could figure out how they got to some of the remote areas!

To find out, researchers captured some crocodiles in the North Kennedy River area. They attached **satellite transmitters** to them. Then they began to track their movements from the river out into the open ocean. They tracked one crocodile from the North Kennedy River all the way up the coast of the Cape York Peninsula and around the tip to the west side. The journey took 25 days and covered 365 miles (587 km).

Surfing Salties

Although salties live in water, they are not great swimmers. To travel, crocodiles "surf" on ocean currents. They can do this for hundreds of miles. The crocodiles only set out on these trips when the tide is changing, so they can "catch a wave."

Tagging saltwater crocodiles allowed scientists to track their ocean crossings.

Warmer water because of climate change may mean salties will migrate even farther in search of cooler temperatures.

Zebra Mussel

Habitat	Native to the Black Sea in Europe/Asia
Known for	Rapidly multiplying and colonizing
Journey type	Invasive species

Zebra mussels spread from Lake Michigan to the Mississippi River basin through a canal connecting rivers from the two areas.

Stowaway Mussels

Zebra mussels can live in different environments in the water. They can even survive out of water for up to seven days. This has helped them become an invasive species in many areas. Zebra mussels may have been accidentally transported to new places in ballast water or by clinging to the bottoms of ships.

The St. Lawrence Seaway in Canada is a busy shipping route from the Atlantic Ocean into the Great Lakes. In 1988, zebra mussels were first spotted in Lake St. Clair, between Lake Erie and Lake Huron. They spread quickly. Today, they are found in more than 750 lakes in North America. They have also spread to Great Britain, Ireland, Italy, Spain, and Sweden.

Survivor

One female mussel can lay up to 1 million eggs each year. Because of their huge population, zebra mussels have a big impact on the lakes and rivers they colonize. They crowd together so thickly that they use up all the food sources.They also block **spawning grounds** for fish. The mussels clog **intake pipes** for power and water treatment plants. Governments spend millions of dollars keeping the pipes clear of the invaders.

Within three months, Zebra mussels can completely cover a hard surface with a colony 0.5 inches (1.3 cm) thick.

Red Crossbill

Habitat	Conifer forests across North America
Known for	Irruption to find food for second breeding season
Journey type	Irruption

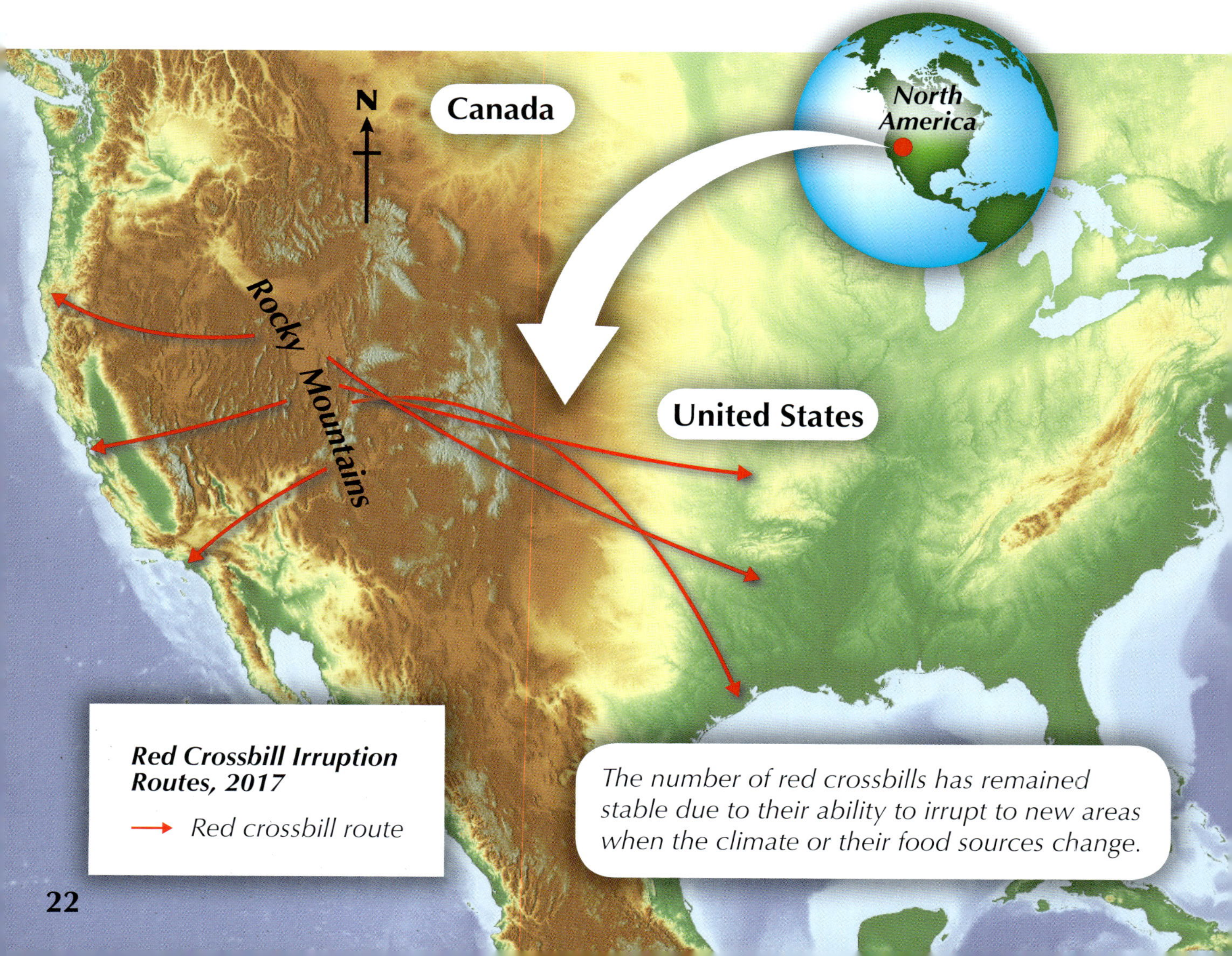

Red Crossbill Irruption Routes, 2017

Red crossbill route

The number of red crossbills has remained stable due to their ability to irrupt to new areas when the climate or their food sources change.

Crossbills call to each other to relay information about whether a tree has a good seed supply or not.

Following the Food

Red crossbills use their scissorlike bill to get to the seeds inside **evergreen** cones. These birds live wherever there are lots of **conifers**, including the Rocky Mountains and across Canada. Unlike most birds, red crossbills breed twice a year. Because of this, they need a good food supply for their chicks all year round.

Cone Crops

Red crossbills irrupt whenever the cone crop fails, such as when there is too little rain or there are forest fires. The most recent irruption was in the winter of 2017 to 2018. The cone crop in the mountains in the west was poor.

Meanwhile, in the northeast regions of Canada and the U.S., there was a bumper crop of cones on spruce, hemlock, and pine trees. In June and July 2017, crossbills began migrating.

By the fall, large numbers had been spotted throughout South Dakota, Kansas, Oklahoma, and Texas. There, the birds nested and feasted on the cone seeds. At the end of that breeding season, the birds made their way west again.

Crossbills usually hold the cone with their foot while they use their bill to pry the scales of the cone open.

Invasive Insects

Habitat	South America, Asia
Known for	Hitchhiking rides in wood or packing materials
Journey type	Invasive species

Ports and airports are the main entry point for insects hiding in packing material.

Hitching a Ride

Each year, more and more products are traded and shipped around the world. Insects get into crates, boxes, and packing material and go with the goods! And these invasive species are affecting people, pets, and plants in their new homes.

Red imported fire ants made the trip from South America to the U.S. in the 1940s. Adult ants or eggs were probably inside the wood of crates or boxes on ships coming from Brazil. They quickly took over the territory of native fire ants. These ants chew through crops and sting anything that moves as they spread north and west.

Invaders

Asian long-horned beetles and Asian tiger mosquitoes are both native to countries in Asia, including China, Japan, and the Korean Peninsula. The beetles arrived in North America in packing crates. Eggs or even adults would be hard to spot in the wood. Asian tiger mosquitoes have migrated among items shipped from Southeast Asia. Some have been traced to the water inside rubber tires arriving by ship. Governments are trying to reduce **infestations** by insisting that wood for crates is treated to repel pests.

Red imported fire ants are very aggressive and will quickly swarm and sting anything that comes near their nests or food sources.

The only way to control the invasion of Asian long-horned beetles is to cut down infected trees and burn or chop up the wood to kill the insects inside.

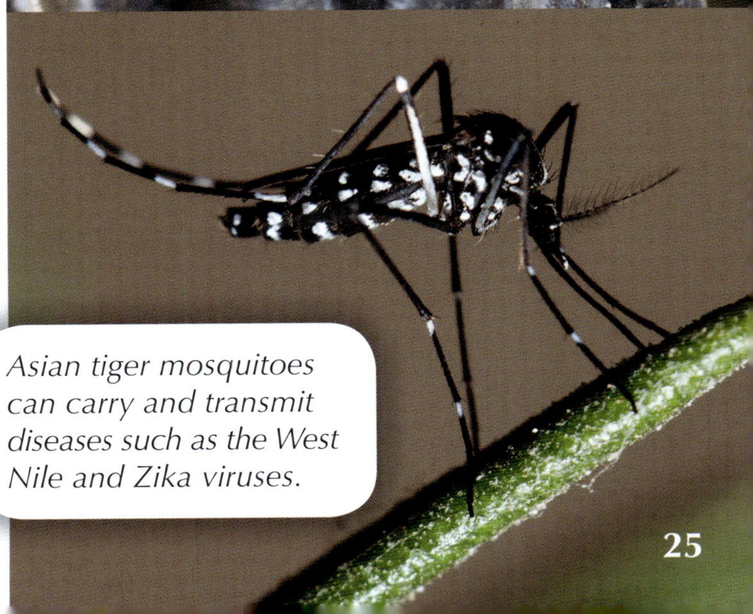

Asian tiger mosquitoes can carry and transmit diseases such as the West Nile and Zika viruses.

New Patterns of Vagrancy

Finding New Routes

Our world is changing. Animal journeys are changing too. Warmer temperatures, severe storms, and expanding cities affect animals. In order to survive, they are traveling into regions they've never been before. Melting sea ice is opening up new areas in the Arctic and Antarctic. This can allow new migration from the Atlantic Ocean to the Pacific Ocean. Gray whales live in the Pacific Ocean. They usually spend summers off the coasts of Alaska and Russia. In the winter, they migrate south to the coasts of California and Mexico. In 2010, a gray whale was spotted in the Mediterranean Sea. It may have crossed through the Arctic as sea ice melted.

Gray whales migrate close to the coast and sometimes swim close enough to boats that the people on board can touch them.

Chasing Food

Birds that rely on fish for food follow their meals across Arctic and Antarctic waters between oceans. Northern gannets eat mackerel and herring in northern Atlantic waters. In 2013, a northern gannet was seen off the coast of San Francisco, California. These birds do not fly over land, so its route must have been over water. As climate change speeds up the melting of sea ice, more animals are changing their breeding, feeding, and migrating areas.

Northern gannets can glide just above the waves for hours, barely moving their wings.

With a strong wind, dragonflies can reach speeds of up to 75 miles per hour (120 kph).

Staying Warm

Warmer temperatures caused by climate change are affecting animals and insects that live on land too. Scarlet darter dragonflies rely on the air to keep their bodies at a constant temperature. As northern areas become warmer, these insects are flying farther north. Red-veined darter dragonflies have been spotted in England and Ireland more often as temperatures get warmer.

Conclusion

Climate change is causing new weather patterns, warmer ocean temperatures, and changing winds. This affects how and where animals move around to feed or reproduce. It can have far-reaching effects for the animals that end up in a new region. Some **thrive**. They colonize a new area, finding new food sources and breeding grounds. Animals that lose one habitat may be able to rebuild somewhere else.

Laughing gulls live in North and South America but have been spotted in Wales in the United Kingdom.

New Neighbors

But often, these unusual journeys have negative results. Individuals may not survive in the new area. If they do, they may not find a mate and raise young. They might also affect the animals that already live there. Invasive species are competition for limited food sources. If they do not have natural predators, their population can explode. This can wipe out native animals. They can also carry diseases which affect both animals and people.

Wild pigs damage fields and crops and can spread bacteria such as E. coli.

Cleaning boats before moving them to a new water source helps prevent invasive species such as zebra mussels from hitchhiking to a new habitat.

Changing Our Ways

There are some ways we can all help slow or stop unusual animal journeys:

- Follow guidelines on what to bring back as souvenirs when traveling. That way, you will not accidentally bring home an invasive species. Things like sand from beaches, rocks, or plant material can have hidden insects in them.
- Report any animals found where they shouldn't be. If there is a living creature in packing material or in fruits and vegetables, quick action can stop them from becoming an invasive species.
- Protect the environment. By preventing pollution, migrating animals have access to clean water and food. We can protect nature so as not to disturb landmarks that birds and animals use to find their way.
- Reduce, reuse, and recycle. This will help slow climate change.

Glossary

ballast water Water placed in the bottom of ships to keep it steady in high waves

breeding grounds Areas where animals give birth to their young

climate change Changing weather patterns over time

conifers Trees that have needles and cones

debris The pieces that are left after something has been destroyed

ecosystem A community of living things

evergreen Trees that keep their leaves all year round

habitat The natural home of a plant or animal

infestation A large number of animals causing damage or disease

intake pipes Pipes where water enters a building or factory

nymph An immature form of an insect

plague An unusually large number of insects or animals causing damage

prey An animal that is hunted and killed for food

satellite transmitters Small devices that send information back to researchers

spawning grounds Places where animals like fish or frogs go to lay eggs

thrive To grow strong and healthy

tundra A flat, treeless area in the Arctic

Find Out More

Books

Kalman, Bobbie. *Invasive Animal Species*. Crabtree Publishing, 2016.

Latta, Sara. *Keep Out!: Invasive Species*. Capstone, 2013.

Markle, Sandra. *Snowy Owl Invasion!: Tracking an Unusual Migration*. Millbrook Press, 2018.

Websites

https://bit.ly/2YkbD7M
From jellyfish to crocodiles, discover some unusual migrations.

www.projectsnowstorm.org
Learn about snowy owls and their migrations with Project SNOWstorm.

https://bit.ly/2KJFmE8
Find out why vagrant birds go out of range.

Index